NAVIGATING THROUGH 21ˢᵀ CENTURY TECHNOLOGY

A Spiritual Warning System

Dr. Derek Hagland

A GRACE BAPTIST COLLEGE PUBLICATION

ISBN: 978-1-60208-339-4

1665 W. M 32
Gaylord, MI 49735
989.732.8881
www.gracebaptistcollege.com

SPECIAL THANKS

First of all, I would like to thank my Lord and Savior Jesus Christ. I so much want to live John 3:30 for Your honor and glory; *"He must increase, but I must decrease."*

I would like to thank Mrs. Debbie Goldsborough and my wife Amanda, a.k.a. "The Hackers", for reading and proofing my work. Words cannot describe how grateful I am for your constant help and many hours of sacrifice in making this book possible. Thank you for assisting me in relating what was in my heart onto the printed page.

I would also like to express my appreciation to Tim Bish. Thank you for the design, layout, and research of all the technological information contained in this book. Your patience in executing all my desired details and changes throughout this entire process was exceptional. It has been a fun journey serving God with you. I look forward to our next project.

DEDICATION

I dedicate this book to my five children: Luke, Bethany, Andrea, Marianne, and Jack. One day you will not have Mom and Dad to help guide and protect you in your use of technology. It is my hope and prayer that the truths in the following pages will guide you in the midst of a world that would like to destroy everything we have taught you.

I love you!

Dad

TABLE OF CONTENTS

FOREWORD

In the age in which we live, technology is both a blessing and a burden. We have all heard the familiar colloquialism 'Can't live with it; can't live without it.' This well-worn quote certainly applies to technology. The genie is out of the bottle. We can rail against technology, but there is no turning back.

Dr. Hagland has tackled this tough issue head on. Rather than bemoan and berate the use of technology, he has thoroughly and thoughtfully given us a roadmap through this tricky and sticky world. The question 'Should we use the internet?' is still a major issue in some circles of Christianity. As a former youth pastor and now a college vice president, Dr. Hagland, rather than giving up or giving in, has hit the nail on the head with this roadmap through the trappings of connected media. The issue is not whether we should use technology; the issue is how we can use it safely. Without a plan, a purpose, and a path with clear and concise guidelines, the use of technology can be spiritually, relationally, and morally dangerous.

We here at Grace Baptist Ministries have chosen to follow the scripturally sound safeguards laid out in this book to help us navigate safely through 21st century technology. I have often said that no Christian can safely use unfiltered internet and media. None of us dare operate free of accountability. To do so is to flirt with disaster and to tempt fate.

Dr. Hagland, thank you for dealing with this complicated and controversial subject with balance and a Biblical approach. This book should be put into the hands of every parent, preacher, principal, and child of God, who, like Daniel, has purposed in his or her heart not to be defiled. I have done more than just endorse the truths of this book – I have embraced them as my own guidelines for navigating safely through 21st century technology.

Dr. Jon M. Jenkins
Pastor, Grace Baptist Church
President, Grace Baptist College

WHY?
Preface

If you were to analyze crime in America, you would learn quickly that in many cases, a common denominator is alcohol consumption. Alcohol abuse is a factor in 40% of violent crimes committed in the United States. This staggering statistic does not account for crimes committed where the consumption was within the legal limit.[1] Alcohol may not be the main factor in the crime, however; it is usually somewhere in the background.

This same phenomenon is true in the areas of technology and moral failure. When students are disciplined in Christian schools and colleges, when marriages are broken, when preachers are dismissed from their ministerial duties, when businessmen are relieved of positions due to moral failures,

---

1. http://alcoholism.about.com/cs/costs/a/aa980415.html

somewhere in the background lurks a common denominator – the sinful use of technology.

Stand before a crowd of teenagers and adults and ask this question. "How many know of a friend, married couple, minister, or business person who has ruined his life or shamed his name and testimony through the improper and immoral use of technology?" Without a doubt, most hands will be raised. What is the problem?

> **273.8 MILLION AMERICANS WERE INTERNET CONNECTED IN 2012.**
> - INTERNETWORLDSTATS.COM

The heart of the problem is that technology has far surpassed our character level. Accessing ungodly things has become incredibly simple. In days gone by, the fulfillment of sinful and immoral desires required great effort and planning. Today is much different.

These desires can be fulfilled without leaving home and with very little effort. Mix curiosity, fleshly desires, and a weak moment, and the mind of a Christian can be instantly soiled by a simple touch of a keyboard.

So, where do we go from here? When it comes to technology, opinions abound. Some declare Facebook a sin, the internet the avenue to all evil, and the smart device a wicked tool – each a means of access to all things immoral. With no safeguards, supervision, and principles, this can very well be true. Do we deal with our digital age by terminating all use of technology, social media, and the internet? Although this extreme would certainly serve to eliminate the temptations that accompany the electronics, it is not practical in our world where technology and its inventions have become interwoven into the very fabric of our culture. Where would we even begin? If the social media began with the invention of the telephone, do we go so far as to eliminate phone usage? No, of course not. The answer to this dilemma is more than a list of do's and don'ts – it is a way of life.

The Christian is to live by Biblical principle. The lists of what to do and what not to do are byproducts of a life lived according to the Bible. If we do not violate Scripture, we can use any electronic device for the glory of God. This truth applies to digital devices now available as well as that which has not yet been invented.

Does God hate technology, or is He simply frustrated with how His people have misused this incredible invention? Can a Christian use technology and still keep his testimony for the Lord?

In this study, **Navigating Through 21st Century Technology**, you will learn practical truths from the Word of God that will reveal where the battle actually takes place and will instruct on how to guard against the dangers that are just a keystroke away.

The misuse of electronics has left a trail of broken hearts, the hearts of both those who have sinned and those who have been affected by the sin. Enough is enough! In the pages to follow, you will find clear direction on how to avoid the traps and pitfalls that are ever present when using technology. The sole purpose of this book is to help the reader honor the Lord with the use

of electronics and one day hear,

Well done, thou good and faithful servant. – Matthew 25:21

May God bless you as you attempt to live a holy life, shining brightly for the Lord Jesus Christ in the midst of a crooked and perverse nation! (Philippians 2:15)

Navigating 21st Century Technology being preached at
the Annual Teen Spectacular hosted by Grace Baptist Church.

NAVIGATE
Introduction

On June 14, 2008, sixteen year old Zac Sunderland launched his sailboat, the Intrepid, into the Pacific Ocean with the goal of becoming the youngest

person to sail solo around the world. This task, which he successfully accomplished, included several rendezvous points where he would meet up with his father and replenish his supplies – points such as Hawaii,

Zac Sunderland and his boat the *Intrepid.*

Australia, and South America. Over the course of his journey, he navigated 25,000 nautical miles at an average speed of six knots per hour.[1]

---

1. http://sports.espn.go.com/espn/page2/story?id=4233223

His experiences were many and varied. He was chased by pirates, and clinging to his pistol he eluded them while calling for help and waiting for that help to arrive. Using his satellite phone, Zac sent emails and hosted a blog which received much attention from those desiring to follow his progress and adventures. Someone from NASA stumbled upon Zac's blog and thought it would be a nice distraction for the astronauts on the international space station to communicate with him.

> **78.6% OF ALL AMERICANS HAVE DAILY ACCESS TO THE INTERNET.**
> - INTERNETWORLDSTATS.COM

A phone call was made from the station and the two navigators had a fantastic conversation, one in the middle of the ocean and the other 250 miles above the earth's surface. After 20,000 miles this young sailor's radar, his life line which allowed him to avoid approaching ocean liners and storms, quit working properly. The absence of a functioning radar meant he had to wake every 20 minutes to see if any ships were in the area.

One evening a few days after the radar went out, Zac was standing near the base of his boat's mast when he heard a low rumbling sound which grew in volume. He grabbed onto the mast and hung on tightly as his vessel was struck from behind by a 30 foot rogue wave. Water poured into his cabin and shorted out his electrical system. Unable to charge his phone, he found himself isolated for several weeks having no ability to communicate with the outside world. During this time he was surrounded by fog that prevented him from seeing more than 50 yards ahead and behind making his world no larger than a football field. His only company was the sound of a slight breeze and the water.

He describes his condition at this time as being "the looniest boy in the world." While telling of his adventures after arriving home safely, he made this powerful and famous statement, "I'd rather be caught in a storm than caught in the doldrums."

Zac Sunderland's story is incredible. Consider a sixteen year old boy you know personally. Now envision him launching out into the Pacific Ocean with the goal of sailing around the world solo. Imagine his parents admonishing him to be careful and to proceed with caution. This whole idea sounds extremely foolish. So what made Zac different from the average sixteen year olds you and I know? The difference is that Zac grew up around the water. His father, a boat builder by trade, taught him the skills needed to navigate on the sea successfully. He was not a novice when it came to experiencing and understanding the demands, dangers, and skills needed to operate a vessel in the ocean.

Why tell Zac's story? His story is important because it illustrates the incredible accomplishments that can be realized in the midst of both known and unknown dangers. It proves that when someone has learned and mastered the proper skills, that person is able to navigate safely around things that could have been potentially life threatening.

When it comes to technology, many people stand on the shore of the vast ocean and prepare to launch out into the dangerous unknown with the simple instructions to "be careful" and "proceed with caution." Taking an average teenager, putting him in a boat with the goal of sailing around the world, and wishing him "good luck" is absolutely absurd. If the only instruction provided was the boat, maps, equipment, and things not to use, this person would fail. Failure would not be due to a lack of ability, but rather a lack of proper guidance in what is needed to ensure a successful journey. If our instructions consisted only of generalities on avoiding storms, pirates, and dangerous situations, we would have little confidence in the leadership provided for getting from point A to point B safely. When it comes to the dangerous seas of technology, giving it our "best try", "being careful", or "proceeding with caution" is a sure recipe for disaster. The Christian must realize where he is, where God wants him to go, and what Biblical skills are needed to reach that destination. Anything less is as foolish as a sixteen year old attempting to solo circumnavigate the world in a sailboat having little or no understanding of the ocean or sailing.

The goal of life for the Christian should be to hear "well done" from our Heavenly Father; everything else should be a by-product of this goal. Attempting to travel through life as a Christian desiring to please God in the use of technology, but with no idea of how to achieve that goal, will set us on a course of disappointment and failure by default. We need spiritual guidelines, safety tips, maps, and navigational equipment to help guide us along our way as we attempt to honor Jesus Christ and strive to hear those wonderful words from our Creator.

What does it mean to navigate through 21st century technology? It means we must spiritually equip ourselves with the tools and skills needed to avoid the destruction that the misuse of technology can bring when we are blown off our divine course. It means we must decide to place our way of living at a higher level. It means we must determine to set our affection on things above. (Colossians 3:2) In everything we do, especially in the arena of the digital age, we are to be people who attempt to spiritually navigate in a world that does not love our Christ, His Word, and our stand for Him. Our standard for conduct must be, "Does this please or displease Christ?" This is the very foundational cornerstone for attempting to honor God when it comes to the use of electronic devices.

There are very few second chances when attempting to accomplish what Zac Sunderland set out to do. The same is true when it comes to setting sail on the scary but exciting waters of technology. Learning to navigate is a must!

14.3 TRILLION WEB PAGES ARE LIVE ON THE INTERNET.
- BUSINESSINSIDER.COM

KEEP
Spiritual Security System

A careful consideration of the instructions and truths of the Bible reveals that every one of us has been commanded by God to guard with all diligence a place of great importance.   To carry out this charge, we must surround this place with a spiritual force field, protecting it from the destructive influence of the enemy, realizing that penetration by ugodly influences will bring about absolute ruin. A breach of this force field will cause the Christian to veer off course, to stray out of God's will, to partake of things that displease Jesus Christ, and to live a life contrary to Scripture. This place that is to be kept (guarded) safe is the heart.

Keep thy heart with all diligence; for out of it are the issues of life.
– Proverbs 4:23

The heart is the innermost seat of a person's being. Many times when the Bible speaks about the heart, the reference is to the mind, the place where decisions are made. The Word of God has much to say about the heart. Through Scripture we learn that the heart of mankind is deceitful and desperately wicked and that it can't be truly known. It can be broken, healed, hardened, made merry, stolen, turned away, given, and cleansed. We also learn that all of the issues of life flow from the heart. As water gushes from a stream creating a path, so does every decision flow out of the heart determining a direction. Each decision made in the heart will propel a person to a set course of action. This truth makes the heart a very important place.

> **300 MILLION PHOTOS ARE ADDED TO FACEBOOK EACH DAY.**
> - ROYAL.PINGDOM.COM

As Christians, we are instructed to keep our hearts. We are to protect the heart by setting up a spiritual guard post. The job of this security system is to watch for anything that may bring harm to the mind, anything which could influence the mind to make choices that would displease God. Realize, however, that the heart is not just to be kept; it is to be kept with all diligence. In other words, we are to guard the heart above all other things with the utmost attention. The implication is that we should guard our hearts not with a cheap lock set from the local hardware store, but with the most elite and elaborate security system available. When we consider all of the things in our lives that need protection, our hearts should be at the top of the list.

If the heart of our nation could be narrowed down to a physical location, that place would undoubtedly be 1600 Pennsylvania Avenue. The White House is where the President of the United States and his family reside. It is home to the office of the most powerful person in the world. It is where decisions are made that affect millions of lives. It is a place where meetings are held that will determine the future course and action of our great country. It is also one of the most guarded pieces of real estate in the world. Consider an article from ABC News concerning the protection of the First Family.

> The White House is more than the first family's home. To the Secret Service: it's a fortress. The iron fence is the first line of defense. Guard stations control the entrances, while bullet-resistant windows protect the occupants. Fence jumpers are not uncommon, so inside the fences are what agents call "perimeters of protection." Alarms are positioned beneath the ground and infrared sensors above the ground to detect intruders. Circulating around the lawns and gardens, often hidden, are groups of armed agents formed into emergency response teams. Their job is to rush forward, not wait for intruders to reach their zone. The Secret Service won't say how many agents there are. They carry semiautomatic pistols, shotguns and machine guns. On the White House roof, teams of snipers keep watch. The Secret Service says they are the best in the world and must qualify every month hitting targets accurately at 1,000 yards.[1]

The White House and its premises are guarded by an elaborate system of alarms. This alarm system, along with the secret service, park police, and Washington D.C. metro police, has one common goal, protecting the White

---

1. abcnews.go.com/WNTstory?id=131253&page=1

House. The guarding of 1600 Pennsylvania Avenue exemplifies the idea of guarding with all diligence. I can think of many businesses and properties that are protected and protected quite well, but when I think of guarding a place with all diligence, the White House comes to mind. Could you imagine giving the White House the same degree of security as a local auto parts store in your own community? Although the local store may have a security system and good measures in place for protection, these would be insufficient for a place as important as the White House. Anything less than a security system that epitomizes all diligence would put our nation, the president, and all who work at the White House at risk. The same is true when it comes to our hearts. Anything less than keeping our hearts with all diligence will place our lives at spiritual risk. As Christians who desire to honor and please God, we must put a concentrated effort into the protection of our seats of decision, our hearts.

---

1 OUT OF EVERY 3 INTERNET SEARCHES IS FOR PORNOGRAPHIC MATERIAL.
- BUSINESSINSIDER.COM

ACCESS
View

For the past several years, Grace Baptist College has hosted an opening week activity at the famous Mackinac Island, a 3.8 square mile island and resort area located in Lake Huron between Michigan's Upper and Lower Peninsulas. Mackinac Island is home to 500-600 people and has an estimated one million visitors every year. Known for its historical significance, landmarks, and natural beauty, the island is also unique in its ban on motor vehicles. The only means of transportation allowed on the island itself are bike or horse, and visitors to Mackinac have only two options to get there: plane or ferry. The island has an airstrip, just large enough to be accessible to smaller type aircraft, and ferry lines which provide transportation over the water from each of the two peninsulas. Unlike Canada's St. Joseph's

Island and other islands which are close enough to the mainland for a bridge, Mackinac Island is located too far from shore. To gain access to Mackinac Island, there are only two choices.

Just as Mackinac Island can be reached two different ways, our hearts

Mackinac Island and the Grand Hotel

have two points of entry: the eye gate and the ear gate. To properly guard the heart from the influences of the enemy, we must understand how these influences gain access and monitor these passageways diligently to keep them free from evil. The Christian who fails to protect these avenues in regards to technology suffers irreparable damage to the mind as it is filled with thoughts and images leading him to struggle spiritually and to fall away from God's plan for his life.

We live in a day of all-out spiritual assault on the heart of the believer, an assault that has always existed to some degree, but has never been so evident as during the present technological age. Through technology the enemy of God has found his way into the hearts and lives of millions upon millions of Christians. The combination of temptation and availability is dangerous for all, believer and non-believer alike. With the simple push of a button or click of a mouse, the devil and every demon of hell can be granted quick access to the heart. The simplest temptation and a weak moment is all our sinful natures need to be compelled by means of technology to indulge in the most depraved things imaginable. We live in a scary day when access to our hearts means opportunity to influence our lives! The wrong things we allow in our minds will influence the decisions we make and the direction we take. Even those who are not naturally bent towards being wicked can by mere curiosity allow evil to have access to their hearts.

The recent electronic "explosion" and the ever increasing ways to be connected and to share information have opened the eye and ear gates of our nation allowing anything and everything to pass through them. The result of our fetish to devour all things electronic is a world that is quickly becoming like the world of Noah. Consider the words of Scripture which describe the times preceding the great worldwide flood, God's judgment on the world:

> *And GOD saw that the wickedness of man was great in the earth, and that every imagination of the thoughts of his heart was only evil continually.* – Genesis 6:5

I can't help but compare what this verse says with the modern day. This seems to adequately describe the 21st century and what is available through means of technology and the internet – great wickedness, the display of every imagination an evil, sinful heart can create via the World Wide Web. Allow me to take this opportunity to remind you of what Jesus Christ said as He described the spiritual condition of the world upon His return.

As the days of Noah were, so shall also the coming of the Son of man be. – Matthew 24:37

It's here! The second coming of Christ draws nigh. He is coming and coming soon. If we are to please Him while living in a day much like the time of Noah, then we must plan on living holy lives remembering that we will be holy not by accident, but on purpose. Purposing to be holy in the world of technology is keeping the heart with all diligence by spiritually protecting the gates that provide access to it. Before they can enter the eye and ear gates, all things wicked must be stopped with a bright neon sign that reads, "Access Denied!"

> **GOOGLE PERFORMS MORE THAN 35 BILLION SEARCHES EACH MONTH.**
> - TECHFEEL.IN

DANGER
Why Leaders Are Frustrated

Our church and college are located in beautiful Northern Michigan, an incredible outdoor recreation and vacation resort venue. No matter the season, Northern Michigan has much to offer – world class golf courses; state and local parks for camping, hunting, and fishing; water recreation on both inland waters and the Great Lakes; and popular ski resorts for snowy winter getaways. Year-round tourism is essential to our economy, and because of this, our natural resources are heavily protected. If you were to visit our college campus, you would see that we are located across the street from the local branch of the Department of Natural Resources (DNR). The entrance to this DNR complex is marked with a sign informing people of the current "fire danger" to the area – low, moderate, high, or extremely high. Although the use of fire always carries a potential danger, sometimes the risk is higher than at other times. In our area, after the snow melts and the ground dries, the summer drought season comes around and the risk of any burning getting out of control and causing a large forest fire becomes a scary reality. Despite the potential danger, however, even during high risk times the DNR will often issue "burning permits" allowing burning as long as all necessary precautions are taken.

Just as there is always the potential of starting a forest fire when open burning, there is always a potential danger with the use of technology, no matter how careful and innocent the user may be. Even the best filtering software is not foolproof – it may malfunction. Just as it takes only one spark to set a fire that could destroy thousands of beautiful acres, misuse of my iPhone, a little 4.5 by 2.25 inch box, could ruin my marriage, bring shame to my personal testimony, cause me to be relieved of my duties as Vice President of our Bible college, and alter God's plan for my life in a matter of seconds!

THERE ARE MORE THAN 18 BILLION
MOBILE DEVICES IN THE WORLD TODAY.
- TECHFEEL.IN

As fire can consume a forest, improper use of electronics can consume our testimony for Christ. The raging fire brought about through the abuse of electronics is one we should seek to avoid at all costs.

> *Be sober, be vigilant; because your adversary the devil, as a roaring lion, walketh about, seeking whom he may devour: Whom resist stedfast in the faith, knowing that the same afflictions are accomplished in your brethren that are in the world.*
> – I Peter 5:8-9

Have you ever seen an area recently destroyed by a forest fire? Every living thing has been burned, consumed. The devil wants to hurt you; he wants to consume you, he wants to devour you! All he needs is an entrance. Don't ever forget that we must always be on high alert to the satanic danger that comes with use of technology.

I am clearly aware of both the pros and cons of today's digital world. I understand the frustrations and the problems posed by technology – in particular, the social media. Through this means of technology, churches, ministries, good families, and individuals have been destroyed. As mentioned earlier, if I were in a crowd at a youth conference, revival meeting, or even an average church and asked how many people knew of someone, whether teenager, adult, businessman, school teacher, or preacher, who has been in major trouble through the sinful use of technology, the number of raised hands would be staggering. That thought alone should help us realize the incredible threat technology poses to our Christian life and testimony.

People hold many differing and passionate views about today's technological world. My viewpoint presented in this booklet is in no way an attempt to make light of the dangers of technology. I realize that the threats and potential dangers are real. Multitudes struggle in this area because they do not know or they choose to ignore the spiritual boundaries which should be in place. It is my desire to present helpful guidelines to aid believers in their service to God.

Christian leaders, including myself, are frustrated with the destruction we see all around us, a result of the use of electronics, the internet, and the social media. The fallout is intense.

One of the major grievances that ministry leaders have with the social media is the provision for unhappy, disgruntled, and unspiritual people to stay connected with others of the congregation. Before social media, when a problem or disagreement occurred in a congregation, the disgruntled person would leave the church or quit coming to youth group, and that would be the end of it. Occasionally there might be contact where the unhappy individual would express his anger and spread criticism, but it was not widespread. Now, through the means of social media, texting, emailing, and such, a disgruntled person has a platform where he can say whatever he wants whenever he wants. Remember, every story has two sides, and every matter should be carefully searched out. Often, however, we are lazy and are content to be headline Christians, easily believing anything that is said to be the truth simply because we saw it posted online.

Imagine a teenager who has gotten angry at a youth director calling the entire youth group together in the teen room and spending hours criticizing the church youth ministry, youth workers, and youth leader. I can't fathom any church leader or member who would think that this meeting and behavior is acceptable. This situation would be Biblical grounds for marking and avoiding this person as one who is attempting to split the church. Consider the words of Scripture:

Now I beseech you, brethren, mark them which cause divisions and offences contrary to the doctrine which ye have learned; and avoid them. – Romans 16:17

Sadly, through the use of technology, this type of unchristian behavior is happening frequently and is damaging the work of God. People are able to spread all kinds of destructive gossip, lies, and criticism while hiding behind a keyboard. I can only guess how many churches and ministries have been destroyed, split, or hindered through this sinful use of technology. No wonder so many leaders seem to be intensely opposed to social media. Through this technology, the whole world has the opportunity to chime in and make comments about what is liked or disliked about a particular ministry and its leader. How frustrating!

> **2.7 BILLION "LIKES" ARE GIVEN ON FACEBOOK EACH DAY.**
> - ROYAL.PINGDOM.COM

Before the invention of the many electronic devices that are so prevalent today, the average person had very few options to access evil. For example, someone with a desire to possess pornographic material had to work incredibly hard and secretly to accomplish that goal. It would require forethought, planning, and a trust between individuals to keep the sin hidden. Through the development of technology and the advancement of the digital world, this kind of evil has become very easy to access. The planning is now limited to a few pushes of a button, the hiding requires minimal knowledge of the workings of the technology, and the result is access to every corner of hell right from one's very own   bedroom! Those who would never dream of frequenting the local adult party store down the street can be lured into this ever present evil through smart devices, the internet, and the world of apps.

Consider the wicked and their habits described in the book of Proverbs.

Enter not into the path of the wicked, and go not in the way of evil men. Avoid it, pass not by it, turn from it, and pass away. For they sleep not, except they have done mischief; and their sleep is taken away, unless they cause some to fall. For they eat the bread of wickedness, and drink the wine of violence.
– Proverbs 4:14-17

Scripture informs us that this craving for wickedness and violence is the appetite of wicked and ungodly people. They feast on the bread of wickedness and the wine of violence, while the gaming, movie, and music industries which feed this hunger are profiting by the billions. Those who satisfy this kind of appetite do so by dining with the devil. If a person is looking for wickedness and violence, the devil has plenty to offer! The disturbing reality in this scenario is that these multi-billion dollar industries are not selling solely to lost, nonreligious people. Many of God's people have pulled up a chair to the devil's table and have been partaking in an "all you can eat" buffet of ungodliness, compliments of Hell. The danger is extremely high, and many Christians seem to be ignoring it, playing with fire. Something needs to change!

It's not enough to simply proceed with caution. Proceeding with caution when it comes to technology can destroy your life by default or by accident. Although some fires are actually started by an arsonist, many forest fires grow as a result of some accident or simple neglect. Many a fire set for a specific purpose and carefully tended has simply gotten out of control and has ended in heartache and devastation. Beware – the fire danger is high when it comes to technology. If we are to proceed, we must do so under strict adherence to the principles found in the Word of God.

As we compare the damage caused by one spark in the creation of a forest fire to the damage done to a person's Christian reputation by one inappropriate use of technology, consider a fire which ravaged our country during the summer of 1910, the largest fire in United States history, a fire known as "The Big Blowup."

The backdrop to this historical tragedy is a series of smaller fires sweeping across the states of Idaho, Montana, and Washington. These fires were attributed to many causes – lightning, land clearing brush fires, and sparks caused from locomotives on the railways. The extremely dry conditions associated with an ongoing drought created the perfect conditions for fires. On August 20, 1910, a bizarre cold front came howling out of the west through the Northern Rockies at 75 miles per hour. These high winds fanned lingering embers and low flames back to life, and many described the next 48 hours saying that the forest exploded. The fire was so large that its smoke reached New England and its soot traveled all the way to Greenland. When it was finally extinguished, the damage was incomprehensible. Over 80 people lost their lives, 3 million acres were burned, an estimated 7.5 billion board feet of timber was consumed, and several small towns were completely destroyed.

This unprecedented national disaster began with a small spark.[1] Beware – the use of electronics can quickly ignite a spark that could ruin a testimony in a matter of moments.

1. idahoforests.org, foresthistory.org, popularmechanics.com

PROCEED
Do & Be

Because we do not want to be among the many Christians who have fallen by means of technology, before we proceed any further, we must consider two things: what we must do and what we must be.

When the fire danger is extremely high, anyone attempting to burn a fire must be well-advised of the laws regarding burning and the penalties for noncompliance. When burning, there is a way to operate which is acceptable, and there is another way which violates the law and results in trouble with the authorities. Before proceeding, cautions must be taken to avoid having a fire that gets out of control.

The same is true for the people of God. If we are going to use technology in a way that does not offend the Lord, then we must understand some things to avoid operating outside of His law. Disregarding man's law is bad enough, but disregarding God's law is even more serious. Before taking another step in the electronic arena, we must understand what we need to do and what kind of people we should be.

> *For the time is come that judgment must begin at the house of God.*
> - I Peter 4:17

> *If my people, which are called by my name, shall humble themselves, and pray, and seek my face, and turn from their wicked ways; then will I hear from heaven, and will forgive their sin, and will heal their land.* - II Chronicles 7:14

What Are We to Do?

We must first judge ourselves and confess our sins. God's people must come clean and get totally real. It is imperative that the whole body of Christ come before God and be honest about our behavior with technology. We must acknowledge the problem. If we are to ever experience God's blessing on our country the way He truly desires to bless us, then we must quit kidding ourselves and deal with this issue on a personal and spiritual level. We must realize more than ever before that to be true followers of Christ, our public and private lives must complement each other, not contradict. The good things we do publicly will never excuse or dismiss those sinful things done privately. The only remedy is to come clean privately as well as publicly. Far too many people have a secret sinful online life. They feel that the sins of their private lives can be ignored because of all the good things they do in public – whether it be teaching a Sunday school class, serving in the church outreach program, or giving generously to a Christian ministry. When we get right with God concerning our private life, then and only then can we truly attempt to please God in all other areas. This is what we must do to navigate properly as Christians living in a technology saturated society.

> *Man shall not live by bread alone, but by every word that proceedeth out of the mouth of God.* - Matthew 4:4b

What Are We to Be?

Jesus instructed us to live our lives according to Scripture – every word that proceedeth out of the mouth of God. We must resolve to be people who live by Biblical principle, people of the Book. Before proceeding any further in the use of electronics, we must determine to be this kind of person. We are to be guided and directed by God's words found in His book. When we stand before the Heavenly Father one day, we will be judged based upon our obedience or disobedience to the Bible. Our Lord has given us clear instructions on how to live our lives. When we proceed within His guidelines, we experience joy, safety, and freedom. Outside of His boundaries we are in spiritual danger.

To navigate spiritually in the technological world, we must be clean and real before God about our weaknesses and our propensity to sin. We must also resolve to live by God's standard, seeking to live according to His instructions, knowing and applying His Word to our use of electronics. Safeguards and godly principles are an absolute must when proceeding within the digital world. If we are to please Jesus Christ in our use of technology, we must both do and be.

PRINCIPLE
Staying on Track

So how do we proceed even though the fire danger is high? To proceed with caution is not enough. The answer is to proceed as God's people have always been instructed, by Biblical principle. Living by Biblical principle is all about staying on track, God's clear path for His people. If we ignore what God's Word teaches, we get off track spiritually, but if we obey His instructions, we keep out of harm's way. Good intentions are not enough to protect us from the temptations and snares that lurk in the world of technology. We need proven guidelines to keep us on course.

In life we can easily get caught up in our day-to-day activities and unintentionally drift off course. This is why it is important for us to dedicate a portion of every day to reading the Bible. The Word of God is filled with principles that will keep us on track, much like the rumble strips found along many of our U.S. highways.

Interstate 75 runs right through our town of Gaylord, and I have traveled it countless times. Years ago, the state installed these rumble strips, also known as sleeper lines, along the outside edges of the highway. If a driver is not paying attention, is being distracted, or is becoming drowsy, these rumble strips serve as a lifesaving warning. When a vehicle drifts over these sleeper lines, the tires make a loud rumbling noise which alerts all in the vehicle that they are veering off track. Only God knows how many accidents these strips have prevented and how many lives they have saved.

Just as the state has installed rumble strips on Interstate 75, we need to install spiritual rumble strips, Biblical principles, into our lives. These principles are truths found in Scripture that warn us  when we are drifting from God's intended path for our lives. These spiritual sleeper lines will help us guard our Christian testimonies and keep us from ruining our lives. Living by these godly truths is a must when it comes to determining guidelines for using technology.

As we learn to live by the principles of God's Word, we obtain a better understanding of how to stay on track spiritually. The main obstacle to staying on track is Biblical ignorance. Knowing the Bible principles is prerequisite to keeping them; we cannot keep principles that we don't know. So many in our churches seem to have a shallow and limited knowledge of God's Word. People know about Scripture, but seem to lack understanding of how to apply it to our twenty-first century culture. This is why, especially as it pertains to technology, we must resolve to become people of the Book, people who both know and practice God's Word. When we learn God's boundaries for living life the way He intended, we discover an incredible world filled with fun and exciting things to enjoy and use for His glory. Applying Biblical truths to our use of technology frees us to do anything with electronics as long as we are not violating Scripture. If something in the technological world requires me to violate Scripture, then I am to proceed no further.

Living by Biblical principle is more than a list of do's and don'ts; it is a way of life. Those who allow themselves to be governed by the truths of the Word of God have set up spiritual rumble strips and guardrails to keep them on the path God desires for all His children. Drifting off His path is both dangerous and displeasing to the LORD. We should seek to know and apply godly principles in our lives with the pure desire to please God in all we do.

Teach me thy way, O LORD, and lead me in a plain path, because of mine enemies
– Psalm 27:11

For over a decade, I had the wonderful opportunity to hold the position of youth pastor at the Grace Baptist Church in Gaylord, Michigan. While I was working with the teenagers, I hosted an annual summer hiking trip in the Porcupine Mountains Wilderness State Park located in the Upper Peninsula of Michigan. This park, seven hours northwest of Gaylord, consists of 58,000 acres (92 square miles). Those of you who have hiked and camped in this type of area realize the importance of keeping to the trails. To leave the main hiking trails is to put oneself in danger. Because the park is so large and the potential so great for a fun-filled nature experience to turn into a life or death situation, the park has taken precautions to ensure that people stay on the trails.

For example, the ranger station provides maps with clear instruction on how to navigate throughout the park. These maps are extremely useful to both first-timers and seasoned hikers as they travel the

thousands of woodland acres. Also, every thirty to forty yards along the trail, the hiker will see a blue diamond-shaped marker lodged in a tree. Sometimes the path is difficult to discern, especially on trails that are seldom traveled or trails where the brush is thick. When the hiker is confused, wondering if he has strayed off the path, these markers serve as confirmation that he is still on track. In addition to these helps provided by the park, as a final safety provision, when the map seems difficult to follow and the markers hard to find, it is best to follow the path that looks well-traveled.

When it comes to technology and living by Biblical principle, I can't help but think back to the techniques and resources we used to keep us on track in the Porcupine Mountains. In life, if we are to stay on track spiritually, we must consult our map, the Bible. Because we don't get a second chance, we need to follow the Master and His plan laid out for us in His Word. At times when the path becomes obscure, we must look for the highlighted markers, Scriptural guidelines that we depend upon to reassure us that we are staying on track. Then, on those rare occasions when we are unsure of the Biblical application in a specific situation and the markers are not obvious, we must travel on the path where good godly men and women have walked before us.

NEARLY 40% OF THE EARTH'S POPULATION GET ONLINE EACH WEEK.
- ITU.INT

Why spend so much time on the issue of Biblical principles and the emphasis to stay on track? Because navigating through twenty-first century technology means knowing how to direct or manage our course. We have all heard about the "Old Time Religion." The old time religion is not something to be afraid of or to disregard simply because it contains the word old. On the other hand, just because something is new does not mean it is bad and an enemy to old time religion. The old time religion has always been about living by Biblical principle in the midst of a wicked world that seeks to live by self-indulgence with no restraint. Those who choose to live according to God's Word are those who can't be bought; their integrity is not for sale at any price. True followers of Christ do what they do because of a desire to obey the

Scriptures and to honor God. The spiritual rumble strips and guardrails are a means to an end. They are placed in our lives to help ensure that we may one day hear, "Well done, thou good and faithful servant."

VIEW
God's Opinion

So, where do we go from here? We have learned of the dangers of technology and the effects its misuse can have on our lives. The dilemma that we face, however, includes the flip side of the issue. Technology brings with it not only a great potential for evil, but also many incredible avenues to spread the gospel and bring glory to God.

Some would declare that the social media and the internet are totally corrupt and that any use of them is sinful. To be consistent with this line of thinking, we must ask ourselves this question. Where do you start? If we are to be consistent and eliminate any technology that keeps us socially connected, do we travel back to the beginning of electronic social media and eliminate telephones? Using technology in a sinful manner is nothing new. Back in the day as the invention and mass installation of home phones brought being connected to a new level, preachers preached against the sin of gossiping over the phone lines. Should we eliminate all technology and live as the Amish do? This option is certainly a safe one and would immediately solve many of the issues with technology. Unfortunately, this option is just not practical in a society that has interwoven technology into the fabric of our everyday lives. The answer is not to eliminate everything. The solution, as stated in the previous segment, is to live by Biblical principle. It's not about eliminating all use of electronics; it's about using electronics without violating the Scripture.

TEENAGERS SPEND AN AVERAGE OF 31 HOURS ONLINE EACH WEEK.
-CSEPEDIA.COM

The sole authority for faith and practice for the believer is the Word of God. Obedience to God's Word settles everything. We ought to read, seek to understand, and apply the Bible to our lives with the intent to live in accordance with the Bible, not in violation of it. The better we know the Scriptures, the more we learn about what God likes and dislikes. Consider what the Bible says about Christ in the book of Hebrews.

Thou hast loved righteousness, and hated iniquity. – Hebrews 1:9a

As Christians, or Christ-followers, we ought to develop the same love for righteousness and hatred for iniquity as our Savior. To do this, we must ask ourselves, "What does God hate and what does He love?" A quick glance at the Bible will reveal that God loves the world (John 3:16), sinners (Romans 5:8), and a cheerful giver (2 Corinthians 9:7). As we dig deeper, we also find that the LORD has given us lists of things He says He hates (Proverbs 6:16-19 and Zechariah 8:17). The Bible does point out in these passages specific things that God loves and hates; however, these lists are not all-inclusive. The Scripture contains much instruction as to what God likes and what He dislikes. As we seek to please God in obedience to the Bible, a basic rule of thumb is this: if it violates Scripture, then it is sin, and if it does not violate Scripture, then it is safe to proceed. The LORD loves obedience and hates disobedience to His Word.

The only opinion about technology that really matters is God's opinion. So, how does God see technology? What is His viewpoint? Is the use of technology sin or can it be used in a way that does not violate His Word? Opinions abound.

While pondering these questions, I can't help but think about all the crazy and passionate product opinions held by people in our American culture. Consider some of the current "either / or" products available on the market: Android or iPhone, Coke or Pepsi, Mac or PC, Xbox or PlayStation, Ford or Chevy, Marvel or DC Comics. As you read this list of choices, I imagine that in some cases you have a personal opinion. You may even know enough about one of these products that your opinion is held with a degree of passion. Passion is good. Let me ask you a question. Have you ever developed a passion for the things God loves and against the things God hates? I am not talking about something as trivial as the temporal; I am talking about the eternal. What will make a difference in light of eternity? The difference is wrapped up in God's opinion and our adherence to it.

Returning to our original question: what is God's opinion on technology? Does God hate technology, or is He more frustrated, displeased, and angry at the way many choose to use it? I tend to agree with the latter. Plenty of technology is used sinfully, but many areas of technology are being used to honor the LORD's Word and work! If we make sure that our conduct lines up with Scripture, we can properly navigate the use of technology and be right with God.

> **CELLPHONE USERS CHECK THEIR PHONES 150 TIMES A DAY!**
> -ABCNEWS.GO.COM

For those who think that all use of modern technology is sin, consider the following ways it is being used for God and for good. Ask yourself: Does God hate this?

1. I think God likes Bible apps, study tools, and Christian sites found online. (As of July 7, 2013, the Bible app, YouVersion, hit 100 million downloads.)

2. I think God likes missions classes at our college when we Skype missionaries from around the world and bring them right to the classroom while conducting facetime sessions.

3. I think God likes the easy access to sermon downloads and Christian music. On our Bible college website (www.gracebaptistcollege.com) anyone can download chapel sermons for free or purchase college tour group music.

4. I think God likes His people easily finding the right church to attend while on vacation. In the past, when a family went on vacation or was away from home and could not attend their normal worship services, finding a church of like faith to attend was a gamble. Now, through the internet, churches can be easily located and identified as to their beliefs. With GPS these churches can be found very quickly and with little hassle.

5. I think God likes the incredible ten hour Genesis to Revelation Bible mini-series that aired in March and April of 2013 on the History Channel. I found it fun to read the write-ups marveling at the enormous viewership of this series which was beating out all the major networks at the time. I thought it was exciting to see the interest our society still has in the Bible and Jesus. It was refreshing to hear all the talk and discussion that took place, even among the lost, as they asked questions because of the broadcast.

6. I think God likes the influence of the movie "The Passion" and the dialog that it created. Regardless of your view on movies that depict the crucifixion, there is no denying that when this movie was released, Calvary became the main topic of discussion among millions of people in churches, homes, and workplaces all over the world.

7. I think God likes His people having access to a world library of Christian literature such as commentaries, resources, Bible dictionaries, and books.

8. I think God likes His children sending encouraging, admonishing, and prayer texts to help people who need uplifting words and intercession.

9. I think God likes us taking prayer requests in our adult "Victory Sunday School Class" and having them typed and sent electronically to all who have signed up to receive our email. So many write down prayer requests and then lose the list or simply forget. This use of email has enhanced our ability to pray for each other in our class specifically.

10. I think God likes the blog where our pastor, Dr. Jon Jenkins, writes instructive and encouraging things that are a spiritual help to all those who have signed up to receive his posts.

11. I think God likes Christian radio stations. One of our church ministries is a Christian radio station with headquarters right on our property. Twenty-four hours a day, seven days a week, 88.1 FM

WLBW is broadcasting conservative Christian music, an hourly presentation of the gospel, several helpful programs, and selective preaching sermons throughout all of Northern Michigan. For more information, visit www.wblwradio.com or download the station app by searching WBLW Radio. Tune in via the web. Listen to Dr. Jon Jenkins' daily broadcast "Echoes from Grace" and my daily broadcast "Reclaim the Name."

12. I think God likes Twitter accounts when they are used to quickly spread information concerning prayer requests, emergencies, and helpful information.

13. I think God likes electronic signatures used in texts and emails that are accompanied by a Bible verse or Scripture reference.

14. I think God likes the personal testimonies in the profile sections of social media pages like Facebook, testimonies declaring a love for Christ and a desire to follow Him. Social media can be a great way to inform others about church services and other special events.

Allow me to add a disclaimer at this time and say that I am not personally endorsing or supporting everything mentioned in this list. I am simply making the point that even though I may not agree with each completely, I am not convinced that God is angry at the use of technology as I have just described.

God's opinion and standard about technology and its use is found in the books of I Corinthians and Colossians. This is God's view:

Whatsoever ye do, do all to the glory of God. – I Corinthians 10:13b

And whatsoever ye do in word or deed, do all in the name of the Lord Jesus. – Colossians 3:17a

FILTERS
Keeping Things Clean

The Transportation Security Administration (TSA) was created after 9/11 to strengthen the security of our nation's transportation system, to ensure "freedom of movement for people and commerce." The TSA is responsible for the screening of commercial airline passengers and baggage, a task which employs 50,000 security officers who screen over 1.8 million people a day.  The screening process requires a conglomeration of behavioral detection officers, explosive detection canine teams, explosion specialists, and security officers.[1] Thanks to the development and diligent work of the TSA, many terrorist attacks and individual plans designed to cause great harm or death to innocent people have been thwarted.

If you have traveled by air recently, you are familiar with the TSA screening process. Proper ID must be viewed along with your passenger ticket. All personal belongings approved as "carry on" must go through an x-ray machine. Shoes and belts must be removed. Computers must be taken out of their travel bags. Only certain size bottles and containers are permitted to pass through the check point. The process is quite amazing, and even though it often seems tedious and inconvenient, this airline security screening is very necessary. In 2012 alone, the TSA detected over 1,500 guns nationwide. This administration must be right every time, it's a matter of life and death. Their screening (filtering) process prevents anything harmful from entering an airport or an airplane.

We have previously learned that all important decisions are made in our hearts (minds) and that the two access points to the heart are the eyes and ears. We also saw that our Scriptural instruction is to guard the heart with all diligence. Just as the TSA screens prevent harmful items from entering airports and planes, a spiritual screen helps extract sinful impurities before they have the opportunity to reach our minds. Spiritual screens help ensure that our use of technology will honor the LORD.

---

1. www.tsa.gov/about-tsa/what-tsa

If we are going to attempt to keep our technology use clean, we must install spiritual filters. Before we go any further, let's talk about filters. What is a filter, and what does it do? A filter is a device used to block impurities, things that contaminate, pollute, or taint. The idea behind installing spiritual filters to help protect us in our use of technology is the extraction of things that could harm and contaminate our hearts and minds. Imagine a spiritual screening process with the same passion and intensity as that of the TSA. Rather than preventing 1,500 guns from getting into an aircraft, this system shields the heart of a Christian from 1,500 temptations that could deeply affect a believer's life and destroy his testimony!

As we consider the screening of the TSA, the process of selecting, considering, and rejecting, by examination, the purpose becomes clear – to filter out anything that is not permitted in air travel. To screen something is to filter it. The function of a filter is to block things that could be harmful. Consider several different kinds of filters that prevent harm or contamination to things we use every day.

The Coffee Filter

Coffee filters hold the coffee grounds and prevent them from contaminating the beverage. Coffee is produced by pouring hot water through the coffee filter and grounds. The filter allows only the water to pass through. No one desires a hot cup of Joe with floating grounds.

The Furnace and Air Conditioner Filter

These filters protect the motor and internal components from airborne particles and dust that may cause damage.

The Oil Filter

A car's oil filter protects the engine by catching particles in the oil which might damage the engine if allowed to circulate.

The Vehicle Air Filter

The exact mixture of fuel and air is essential for proper operation of a vehicle. All of the air first enters the system through an air filter which catches dirt and other foreign particles preventing entry into the system and possible damage to the engine.

In all four of the examples given, a filter protects the function and outcome of something. When it comes to vehicles, furnaces, and air conditioning units, careless use without proper filters could cost thousands in repair bills or even tens of thousands in replacement expenses. A filter is essential to the operation of some very important and expensive machinery. In the same way a spiritual filtering system is essential for the protection of one of God's most complex and incredible creations, the human heart and mind.

**THERE ARE 6.8 BILLION CELLULAR
SUBSCRIPTIONS WORLDWIDE.**
-ITU.INT

For several years while serving as a youth pastor, I was also in charge of the buildings and grounds of our entire ministry. I would frequently have the heating and cooling filters checked and replaced. I would have to investigate, determine how many filters needed to be changed, pick up the phone, and place an order. Due to the size of our ministry, one filter was usually not enough. I would need several filters at a time which required a bulk order purchase. For optimal operation of our system, it was not enough to just know that the filters needed to be changed. I had to make an actual transaction. I needed to buy the filters and have them installed.

Buy the truth, and sell it not; also wisdom, and instruction, and understanding. – Proverbs 23:23

Proverbs 23 instructs us to make some spiritual purchases. As we attempt to live a life pleasing to God, we need to do some inward investigation and make some investments in our Christian life. Just as I had to purposefully place an order, purchase, receive, and install filters for our heating and air units at our church, I also need to purposefully place a bulk order and install several spiritual filters into my life. These six filters, essential to guarding our hearts by the proper use of technology, are found in Philippians chapter four. While attempting to navigate above reproach in our digital age, we can use these scriptural filters to both prohibit and remove things that could spiritually contaminate our lives.

> **OVER 1 MILLION APPS ARE DOWNLOADED EACH HOUR.**
> -GOOGLE.COM

What Happens in an **Internet Minute?**

- 639,800 GB of global IP data transferred
- 6 — New Wikipedia articles published
- 135 — Botnet infections
- 1,300 — New mobile users
- 20 — New victims of identity theft
- 204 million — Emails sent
- 47,000 — App downloads
- 100+ — New LinkedIn accounts
- 583,000 — In sales
- 61,141 — Hours of music
- 20 million — Photo views
- 320+ — New Twitter accounts
- 3,000 — Photo uploads
- 100,000 — New tweets
- 277,000 — Logins
- 6 million — Facebook views
- 2+ million — Search queries
- 30 — Hours of video uploaded
- 1.3 million — Video views

In 2015, it would take you **5 years** to view all video crossing IP networks each **second**

And **Future Growth** is **Staggering**

- **Today,** the number of **networked devices** = the global population
- **By 2015,** the number of **networked devices** = **2x** the global population

Source: Intel Corp

WHAT HAPPENS

ONLINE IN 60 SECONDS

ON THE INTERNET, WE ALL KNOW THINGS CAN MOVE AT A LIGHTNING-FAST PACE. IN JUST A MINUTE, YOU CAN READ THROUGH AND COMPOSE A FEW TWEETS ALONG WITH LOOK AT DOZENS OF FACEBOOK PHOTOS. THAT SAID, WE'VE PULLED TOGETHER THIS INFOGRAPHIC TO GIVE YOU AN UPDATED VIEW OF EVERYTHING THAT HAPPENS ONLINE IN 60 SECONDS DURING 2013.

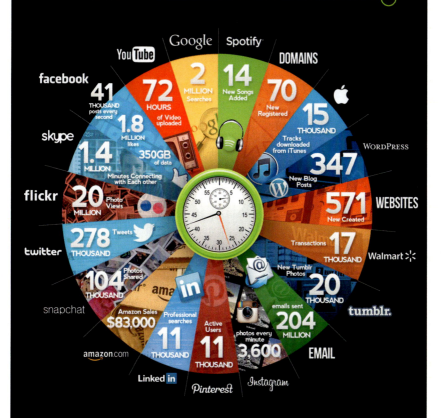

facebook — 41 THOUSAND posts every second

skype — 1.4 MILLION Minutes Connecting with Each other

flickr — 20 MILLION Photo Views

twitter — 278 THOUSAND Tweets

snapchat — 104 THOUSAND Photos Shared

amazon.com — Amazon Sales $83,000

Linked in — 11 THOUSAND Professional searches

Pinterest — 11 THOUSAND Active Users

Instagram — 3,600 photos every minute

You Tube — 72 HOURS of Video uploaded · 1.8 MILLION likes · 350GB of data

Google — 2 MILLION Searches

Spotify — 14 New Songs Added

DOMAINS — 70 New Registered

15 THOUSAND Tracks downloaded from iTunes

WordPress — 347 New Blog Posts

WEBSITES — 571 New Created

Walmart — 17 THOUSAND Transactions

tumblr. — 20 THOUSAND New Tumblr Photos

EMAIL — 204 MILLION emails sent

SOURCES:

PCMAG.COM   GO-GULF.COM   BUSINESSINSIDER.COM   DAILYMAIL.CO.UK
4MAT.COM   SCOOP.INTEL.COM

PHILIPPIANS 4:8
Access Denied

Special events such as concerts, political speeches, and sports competitions often have specific areas that are not available to just anyone. To enter these areas one must obtain what is called an "event pass" or "backstage pass." Going a step further, to have access to everything, one must secure an "All Access Pass." Entrance to VIP areas or places that are off limits to the general public are guarded by security personnel who allow only authorized people to proceed. Everyone else is denied access. This filtering process allows passage to the approved and restricts passage to the unapproved.

During my tenure as the youth pastor of Grace Baptist Church, for many years I hosted our Annual Teen Spectacular Youth Conference. In planning this event, we would choose a theme and attempt to use it to the fullest. One year, in order to film video footage and conference commercials for our NASCAR themed conference, "Ultimate Overdrive," we secured pit passes at Michigan International Speedway (MIS). This pass gave us access to the pits for up close videos of the drivers and cars; however, it did not give us unlimited access. We were not allowed to go anywhere we wanted, and we were not allowed to speak to the drivers. Anything we secured on video was purely by happenstance, a result of being in the right place at the right time.

Those of you who follow this racing sport will probably be familiar with the name Kasey Kahne. Because we had procured a pit pass, we were able to speak with Kasey and shoot an opening conference video of Kasey welcoming all of our delegates to Teen Spectacular.

Another year we did "Cowboy Up," a conference theme based on the sport of Professional Bull Riding (PBR). To attend a PBR event, we were able to secure press passes which gave us access to nearly everything. It was incredible! We had prearranged interviews with famous Christian bull riders, a special video filming box, and the ability to get right up on the platform where the riders mount the bulls while the event is taking place. We were given special stickers to wear on our shirts during the event which allowed us to go backstage, to the media room, and to the prep room where the riders were located. Without these passes, access would have been denied.

I am writing this portion of the book while traveling with the Men of Grace summer tour group out of Grace Baptist College. Just recently while attending the Powerhouse Youth Conference hosted by the Harvest Baptist

Church in Natrona Heights, Pennsylvania, we found ourselves with a free afternoon. Because we were in the Pittsburg area, we thought it would be fun to take some time for an official tour of Heinz Field where the Steelers play. Needless to say, my son Luke, who is an avid Steelers fan (my apologies to all those Packers fans out there) was beside himself. Once again, we received stickers to wear which let all of the workers and security personnel know that we had a pass, permission to be in places where fans are not always allowed to be. Because

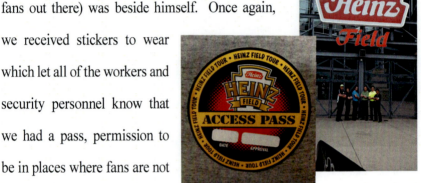

we had these passes, we were able to view the press box, the $100,000 a year suites, and the Steelers locker room. We were even able to walk out of the official entrance tunnel onto the main field. Once again, the pass was the key to being granted access to these places.

Without the pit pass for the NASCAR event (preapproval), we would have been denied access, no questions asked. Without the media passes for the PBR event, the Van Andel Arena staff would never have given us clearance to be backstage and to interview the bull riders. We would have been denied access, no questions asked. If we had simply walked off the street and attempted to enter Heinz Field on our own, we would have been confronted and asked to leave immediately. Access would have been denied.

For far too long Christians have let down their guard and allowed hurtful things to have free access to their hearts and minds. Some things have no business getting an "All Access Pass" to our hearts. Protecting our mind is about restricting access. To discern what should be approved and what should be denied, let's look at the spiritual filtering system found in Philippians 4:8.

Finally, brethren, whatsoever things are true, whatsoever things are honest, whatsoever things are just, whatsoever things are pure, whatsoever things are lovely, whatsoever things are of good report; if there be any virtue, and if there be any praise, think on these things.

This verse serves as a spiritual security guard watching over all that attempts to gain access to our minds. When things that violate this Scripture try to influence us, those things are immediately stamped and labeled as "Access Denied"! The security guards that ought to be constantly protecting our hearts are the six spiritual filters that can be derived from this list found in Philippians 4:8 – things that are true, honest, just, pure, lovely, and of good report. This is the VIP list of things that are allowed access to our hearts. Anything contrary to this list is to be snagged by our spiritual screens and denied access, no questions asked!

Let's take a closer look and consider these Biblical filters that are available for use in our attempt to navigate through 21st century technology. From Philippians 4:8, we can determine what God approves and does not approve. If anything contrary attempts to gain access at the door of our minds, we can stamp it with a big, fat "ACCESS DENIED"!

<u>Contrary to Scripture</u>: **Whatsoever things are true!**

We are instructed to think on things that are true. Truth receives an "All Access: VIP" pass into our hearts and minds. Truth is that which is uncovered or revealed, that in which there is no deceit or falsehood, that which can be considered fact. The complete opposite of truth is a lie or a cover up. Because truth receives approval to influence our lives, anything that does not line up with truth should be denied access and the opportunity of influence. Our goal here is to stop any impurities that are Contrary to Scripture.

The Contrary to Scripture Technology Filter – Stops the following impurities from reaching the heart:

Any idea, philosophy, or teaching that opposes or is opposite to the Bible.

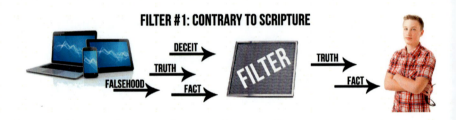

The way this filter is properly and effectively used is through Bible knowledge. Bible knowledge can be obtained through attending youth group meetings, youth conferences, revival meetings, weekly church and Sunday school classes where God's Word is emphasized and taught, as well as daily Scripture reading and personal Bible study. In order to know if something is contrary to truth, we must seek to understand what is true.

<u>Dishonorable:</u> Whatsoever things are honest!

We are instructed to think on things that are honest. Honesty receives an "All Access" VIP pass into our hearts and minds. Honesty is that which is honorable. The complete opposite of honorable is that which is dishonorable. Because honor should receive approval to influence our lives, anything that is dishonorable should be denied this access and opportunity. Our goal then is to stop any impurities that are Dishonorable.

The Dishonorable Technology Filter – Stops the following impurities from reaching the heart:

Anything that hurts a personal testimony for Christ, dishonors the name Christian, or is considered shameful or disgraceful for a child of God.

<u>Unrighteousness:</u> Whatsoever things are just!

We are instructed to think on things that are just. That which is just receives an "All Access: VIP" pass into our hearts and minds. The word *just* refers to things that are righteous, right, or perfect. If something is just, it is considered moral and holy. The complete opposite of just is unjust and unrighteous. Because things that are just receive approval to influence our lives, anything that is unjust should be denied this access and opportunity. Our goal then is to stop any impurities that are Unrighteous.

The Unrighteousness Technology Filter – Stops the following impurities from reaching the heart:

Anything that is obviously sin, ungodly, or has an obvious hellish origination or influence.

<u>Defile:</u> Whatsoever things are pure!

We are instructed to think on things that are pure. That which is pure receives an "All Access: VIP" pass into our hearts and minds. The word *pure* is separate, uncontaminated, and unmixed. If something is pure, it is considered to be free from all "foreign matter." The complete opposite of purity is impurity, that which has been made dirty or unclean, contaminated. Because things that are pure receive approval to influence our lives, anything that is impure should get denied this access and opportunity. Our goal then is to stop any impurities that Defile.

The Defile Technology Filter – Stops the following impurities from reaching the heart:

Anything that influences a person to commit sinful actions and thoughts

that make him filthy before a holy and just God.

Ungodly Desires: Whatsoever things are lovely!

We are instructed to think on things that are lovely. That which is lovely receives an "All Access: VIP" pass into our hearts and minds. The word *lovely* refers to things that are acceptable or pleasant. If something is lovely in God's eyes, then it is considered Scripturally appropriate. The complete opposite of things that are appropriate and acceptable is unacceptable. Because things that are lovely receive approval to influence our lives, anything that is unacceptable should be denied this access and opportunity. Our goal then is to stop any impurities that create Ungodly Desires.

The Ungodly Desires Technology Filter – Stops the following impurities from reaching the heart:

Anything that involves lust, greed, covetousness, and a desire for anything

other than what God desires a Christian to hunger and thirst for.

Mean-Spirited Words: Whatsoever things are of a good report!

We are instructed to think on things that are of good report. That which is of good report receives an "All Access: VIP" pass into our hearts and minds. The words *good report* refer to things spoken with a kindly spirit and good will to others. If something is of good report, it carries the intention of exhorting the hearers and readers. The complete opposites of things that are of good report are reports that are bad, critical, mean-spirited, and evil in nature. Because things that are of good report receive approval to influence our lives, anything that is considered an evil report should be denied this access and opportunity. Our goal then is to stop any impurities that are labeled as Mean-Spirited Words.

The Mean-Spirited Words Technology Filter – Stops the following impurities from reaching the heart:

Anything that involves or has the spirit of being critical, judgmental, vengeful, destructive, malicious, personally hurtful or damaging in nature and character defamation.

FILTER #6: MEAN-SPIRITED WORDS

RESULTS
Determining the Outcome

Have you ever considered Daniel's situation when he was taken captive during the Babylonian captivity? Every time I come to Daniel chapter one, I am inspired by his story. Consider his situation – he is alone, away from his family and friends. His freedom to visit the temple, sacrifice, and receive spiritual instruction from his leaders has been lost. All external pressure to live a holy life before God has been removed. Yet, when presented with the opportunity to partake in things that would violate Scripture and displease God, he chooses to do what is right. Why? Where did he find the strength and ability to do what was right against so many odds?

The answer is found in Daniel 1:8,
"But Daniel purposed in his heart that he would not defile himself with the portion of the king's meat, nor with the wine which he drank: therefore he requested of the prince of the eunuchs that he might not defile himself."

Daniel predetermined the outcome. He had already decided what he would do and how he would respond if faced with a situation that would cause him to disobey God's Word. The battle plan had already been laid out in his mind. He had already purposed in his heart what he would and would not do, and in doing so, he determined that the end result would be one that glorified his God.

Put yourself in Daniel's shoes and apply his situation to living life in the 21st century. You are out on your own away from your parents' authority and leadership. You are not accountable to any church, pastor, or youth pastor. You can go where you want and do what you please. You have the world of technology at your fingertips and you may listen to and view whatever you desire. What will you do and why? What won't you do and why?

Those who have never purposed in their hearts what they will do when faced with temptations are those who don't have a fighting chance spiritually. The opportunity presented provides an easy avenue for many to dive headfirst into the very depths of sin, and sadly, this is happening right before our very eyes. Christian young people get out on their own, purchase or receive as a gift their first smart device, and without proper preparation, unwisely jump into a technological world unprotected from the evils that lurk for the precious and innocent life. In all reality, this scenario is what is waiting for you. One day you will be presented with the opportunity to do whatever you want. If you truly want your life to bring honor and glory to God, you must "purpose in your heart." In other words, you need a plan for your spiritual life. The spiritual filter system is designed for this purpose.

Proper use of the spiritual filtering system will ensure appropriate use of any electronic device. Imagine the Christian lives and testimonies that could be saved or altered if, before stepping out into the crazy world of technology, we first passed everything through the Scriptural screening process. Using this process of Biblical filtering is a way of purposing in our hearts to do right before we are thrust into a spiritual danger zone. This type of cautious operation, living according to Biblical principle, can yield wonderful results for the believer who desires to remain a clean, untainted vessel for the Lord's use.

DETERMINING THE OUTCOME

Far too many people battle daily with things which haunt them, things they have viewed or listened to online. Through Scripture, prayer, and Bible-based counseling, victory can be achieved, but not without residual scars and hardships. I challenge you, skip the utter torment that results from exploring all the world has to offer and guard your heart and mind from the influence and power of our adversary, the devil. He wants nothing more than to completely destroy anything in your life that resembles Jesus Christ, as well as anything you might be able, with God's power, to accomplish for Him and His glory.

Be sober, be vigilant; because your adversary the devil, as a roaring lion, walketh about, seeking whom he may devour: Whom resist stedfast in the faith, knowing that the same afflictions are accomplished in your brethren that are in the world. But the God of all grace, who hath called us unto his eternal glory by Christ Jesus, after that ye have suffered a while, make you perfect, stablish, strengthen, settle you.– I Peter 5:8-10

If Daniel was going to honor God with his life, he was going to have to live differently than others. Remember, there were thousands of young people taken captive, but we only learn of a few. The one who stands out among them all is Daniel. As we study his life and testimony, we find that his decisions at an early age to live by Biblical principles brought about some incredible things – God elevated him to a place of powerful leadership; he was able to stand for God no matter what the cost; he spoke with angels; he was used to pen a portion of God's Word; and he was given great prophetical insight concerning the end times. He was blessed and used of God in a great way all because he purposed in his heart that he would do those things that yield God's favor and abstain from those things that do not. The story of Daniel's life is a testimony of one who *"pressed toward the mark for the high calling of God."*

But what things were gain to me, those I counted loss for Christ. Yea doubtless, and I count all things but loss for the excellency of the knowledge of Christ Jesus my Lord: for whom I have suffered the loss of all things, and do count them but dung, that I may win Christ, And be found in him, not having mine own righteousness, which is of the law, but that which is through the faith of Christ, the righteousness which is of God by faith: That I may know him, and the power of his resurrection, and the fellowship of his sufferings, being made conformable unto his death; If by any means I might attain unto the resurrection of the dead. Not as though I had already attained, either were already perfect: but I follow after, if that I may apprehend that for which also I am apprehended of Christ Jesus. Brethren, I count not myself to have apprehended: but this one thing I do forgetting those things which are behind, and reaching forth unto those things which are before, I press toward the mark for the prize of the high calling of God in Christ Jesus. Let us therefore, as many as be perfect, be thus minded: and if in any thing ye be otherwise minded, God shall reveal even this unto you. – Philippians 3:7-15

To me, there is nothing that better illustrates these verses than the dedication and commitment of those who make the Olympic team and attempt to win the gold. The story of Gabrielle Douglas is a wonderful example of this type of dedication. Gabrielle Douglas left home at age 14 and moved 1200 miles away to live with a host family while she trained for the 2012 Summer Olympics. For nearly two years she was separated from her friends and family. She was homeschooled online so she could dedicate forty hours a week to her gymnastics training and travel extensively to national competitions. While other girls were taking advantage of the many fun and exciting things common to young ladies her age, Gabby was adhering to a strict schedule, special diet, daily training, incredible pressure, and severe homesickness. Back home, her family's financial sacrifice was pushing them into bankruptcy and dependence upon food stamps.

Watching the 2008 Olympics had fueled her ambition to train for the London Games, and her focus was on one thing, winning the gold medal. She did just that and more! At age 16, she emerged as an American icon and role model for millions of others as she won both the gold team medal as part of the American gymnastic team and the gold medal in the gymnastics individual all-around event, the first African-American to do so.

As Gabby returned to the States and she and her family began to reflect on all that she had accomplished, they came to the conclusion that it was worth all the sacrifice. The fame and attention from newspapers, magazines, and major network interviews proved to her that her accomplishments were an inspiration to others, and the many endorsements that came as a result of her sacrifice brought instant wealth.

The fun usually comes long after the sacrifices. It has been said that

---

YOUTUBE USERS WATCH MORE THAN 4 BILLION HOURS OF VIDEO EACH MONTH.
-THESOCIALSKINNY.COM

---

hindsight is twenty/twenty. How true this is! Many times it is only when you are farther down the road looking back in reflection that you can clearly see the benefits of little decisions and choices made long ago. Everyone wants to be a champion, but few are willing to do what it takes. If you want to be a champion for Christ, then you will have to get serious about what it takes to be a champion Christian in God's eyes. That means, when it seems that everyone else is freely and secretly surfing the web and partaking of things that soil a Christian testimony, you, standing on principle, do not. Whether or not you could get away with sinful or questionable technology, you have already purposed in your heart that you will not. You have a goal in mind, a prize – a desire to please the Lord Jesus Christ with your life. If you are going to be a champion for Him, you can't settle to live like everyone else.

With that thought in mind, I want to point out a somewhat famous picture of Gabby. Across the photo are written these words, *16 and a champion, not 16 and pregnant*. While other girls were living life the way they wanted, she decided to make some present sacrifices for future blessings. Did it pay off?

Absolutely! The same is true in the Christian life. If you want your outcome, your future, to be one that honors the God of your salvation, then you must do the same spiritually.

Decide now – choose to do the right thing when it comes to technology so that you too can one day experience what it feels like to be a spiritual champion.

The testimony of far too many is one of what not to do. God give us people who will be a living example of doing right. If you want your outcome to be pleasing to the Lord, then you must purpose some things in your heart. May your picture someday read, Champion for Christ, not Unable to Qualify.

REMINDER
A Spiritual Advisory Label

The writing of this book followed the development of a sermon I prepared for the annual Teen Spectacular Youth Conference hosted by our church each spring. Drawing from several different influences and experiences, after organizing my thoughts, I finished the preparation for the message, but still felt that the process was incomplete. Something was missing. In my thinking, I searched for a way that the message and truths presented would live on long after the oral delivery. While praying and thinking about this, God gave me the idea of the Spiritual Advisory label. It is my hope and prayer that this will serve as a help and constant reminder of the ever-present dangers that accompany technology and all of its gadgets.

The Concept:

Many products such as music CDs, movies, and video games are labeled with warnings and ratings concerning their content. One of the most recognizable logos is the Parental Advisory label. This icon warns parents of extremely explicit language and behavior contained in a product. In our ever-changing

technological world, we must be constantly reminded that although technology brings many fun, educational, and helpful benefits, it can also provide access to videos, language, and relationships that could be spiritually destructive to the life of a 21st century Christian. With this need for constant vigilance in mind, we have created a Spiritual Advisory visual reminder, small sticker labels displaying the letters SA.

The Label:

The new Spiritual Advisory label is designed to serve as a reminder on technological devices. It is a visible, two-letter symbol cautioning the user that technology used apart from Biblical principles has the potential to spiritually ruin a Christian's mind, testimony, and possibly, God's plan for his life.

How to Use:

Take the SA stickers (found with the CD accompanying this book) and place them in visible places on your devices such as car radio, laptop, monitor, home TV and remote, iPad, tablet, iPod touch, and smart phone. Every time you see this sticker, let it remind you that this object could spiritually harm you and thwart God's will for your life.

The Spiritual Advisory Sticker Warning:

Warning: This device has the ability to allow you to see and hear things that could potentially hurt you spiritually. Without proper caution and strict adherence to Biblical principles, the user could be in extreme spiritual danger. DO NOT proceed with caution, but proceed only under Biblical principle and through the proper spiritual filtering system found in Philippians 4:8.

Spiritual Reminder:

This label will not stop anyone from doing wrong, but it can help prevent sinful actions with an electronic device by simply being a visual reminder of potential danger. When God speaks to you at a church service, revival meeting, or conference, you take a trip to the altar and make some solid spiritual decisions and maybe even some good changes to your Christian life. Once you have been stirred to make some spiritual changes in your life, you usually leave these kinds of services with a renewed resolve to follow through on your commitments. After time, you can stray from those decisions. This sticker will help revive your desire to honor God every time you use your technological device. With the use of the SA sticker, you are continually reminded that the object you are using could cause spiritual harm.

Think of all the people who have lost jobs, ruined a marriage, caused untold heartache to their families and friends, hurt the work of God, the name of Jesus Christ, and their personal testimony all because of the inappropriate use of technology. Think of the potential of a simple reminder. Warning! Proceed, but only under Biblical principle. The time for a label such as this is not during the revival meetings we experience, but in the weeks that follow when we are a little tired, vulnerable, and backslidden. That is when we need something to help rekindle us and remind us of those sacred times when God dealt with us and we made deliberate decisions concerning our faith. This reminder label could save your Christian life!

Extras:

Also accompanying this book and the Spiritual Advisory labels is a disc containing the video sermon as preached at Teen Spectacular and an audio version as preached at Grace Baptist Church at Dr. Jon Jenkins' request. These messages have been included to give the reader an opportunity to catch the spirit and heartbeat of the sermon and speaker. In addition, you will find several computer downloads such as The 10 Technology Commandments, 4 Questions to Ask When Using Technology, and more. May God bless!

BEWARE
Application Necessary

So where do we go from here? Like a good, well-studied sermon that is filled with powerful principles and ideas backed up by the truth of God's Word and delivered under the direction of the Holy Spirit, it is only as good to the listener as it is applied in his life. Throughout this book we have taken a journey illustrating how a person can Biblically navigate through 21st century technology. But like a good sermon, this book is valuable to the reader only to the degree that the truths are applied to life.

The Biblical filtering system and the Spiritual Advisory label are only as good as they are applied. They are simply tools to use to accomplish a much higher goal – pleasing God. I love the words that display the testimony and consistency of the example that Jesus Christ left for us to follow; *"...I do always those things that please him."* (John 8:29b) That is the purpose of this book, that while we enjoy the incredible benefits and luxuries of today's technological world, we too would always do those things that please God.

All this has been said in order to offer you this disclaimer: Beware, application necessary! Tools don't build houses, contractors do. To do so, they take the tools they have and use them accordingly. Purchasing heat and air unit filters does absolutely nothing unless they are installed. The examples could go on and on. Just stop to think for a moment about all the things we use in our lives that require application: shampoo, lotion, soap, band aids, shoe shine, car wax, deodorant, toothpaste, pain reliever, and medicine. Just having these things is not enough; to be effective they all must be applied.

Our city of Gaylord is located one hour from two of the Great Lakes. Like anyone, our family loves to spend time in the sun, and we have a favorite spot along Lake Michigan we like to frequent during the summer months. Believe it or not, we will sometimes endure temperatures in the 90s. The difference between our hot days up north and the heat in the south is the humidity. We have almost no humidity. Our heat is dry, and it is very easy to get burned from the sun. Sunburn can be both serious and dangerous.

During some of our trips to the lake, we have made the mistake of putting sunscreen on too late or of forgetting to reapply every two hours. The result is sunburn. The most important thing about protecting yourself with sunscreen is remembering to apply it. Sunscreen does nothing if you just have it; you have to apply it. You can have the best, longest lasting, waterproof sunscreen available, but if it remains unapplied, you will get severely burned. Application is the key. In the same way, if you fail to uphold this principle in your use of technology, the application of the truths presented in this book, you will get burned spiritually.

I love the GPS system on my iPhone! All I have to do is punch in the address of where I want to go, and the GPS will lead me, turn by turn, visually and audibly, to my desired location. There have been times I have not followed the directions, and when I do, the GPS will immediately attempt to reroute me and get me back on course. This book is my attempt to give you the spiritual coordinates to help you arrive at a technological destination that is pleasing to God. Now it's time for you to enter them into your heart and mind and diligently follow the principles as the Holy Spirit talks you through every step of the way. Remember, if you mess up, ask the Lord to forgive you and allow the filtering system of Biblical principles to reroute you back on course.

May the Lord bless you while you seek to honor Him as you spiritually navigate through 21st century technology.

**22 BILLION TEXT MESSAGES
ARE SENT EVERY DAY.**
-ABCNEWS.GO.COM

SILENCE!
The Value of Quiet Time

With all the benefits of technology and its devices, we have found ourselves immersed in a society that is addicted to electronics. That addiction, constantly needing to be connected, always checking on or doing something online, has plunged us into a world of noisy distractions. The noise comes from an electronic age that is constantly screaming for our attention. Because of this attraction to all things electronic, we seem to have lost our sense of the value of quiet time.

Meditation:

One of the values of quiet time is the opportunity to stop and meditate. Meditation, the art of just thinking on things apart from distractions, is a much needed practice in our fast-paced, impulsively acting and reacting society. When we contemplate and consider things, especially spiritual things, we force the noisy distractions to take a backseat to the still small voice of God.

When was the last time you took a few moments to push away from technology so you could do some deliberate thinking about all the important issues you face in life? Maybe a little silence and meditation would allow you to quiet the demands of the world long enough to process the many decisions you need to make. We are all presented with different opportunities, and we all need God's direction as we attempt to follow Him. Quiet time will aid in this process and will help you to clearly think things through.

Devotions:

Another value of quiet time is the opportunity to get away from everything for the purpose of Bible reading and prayer. Far too many of us allow alerts, texts, tweets and such to rob us of much needed fellowship with our Creator. Take a moment to set aside a period of time where you plan to be free from things that distract you while you pray and read your Bible.

Family:

Technology has crept in and is stealing valuable family time. We have all become too complacent with a scenario of everyone being together under the same roof, but the family time consisting of each family member doing their own thing via technology. How sad it is when a family decides to spend time together, participate in an activity, or go on a vacation and that family time is spent secluded in texts, Facebook, video games, emails, and movies.

We all know that the family as a whole is under attack. The enemy has used the incredible inventions of technology to put a wedge between our personal communications which are essential to building influential relationships. Communication is the key to any relationship. We don't talk anymore. We don't take time to know our parents or children. These days will soon pass. Family is important! Let's make time to build strong family relationships and memories in an environment that is occasionally free from electronic diversions.

> **THERE ARE MORE INTERNET CONNECTED DEVICES THAN THERE ARE PEOPLE ON THE EARTH.**
> -THESOCIALSKINNY.COM

Rest:

With so many incredible devices created to make our lives less complicated, has anyone else stopped for a moment and realized how exhausted we are? The easier things seem to get, the more tired we have become. We all need a period of time where we are not at the beck and call of the immediate. No wonder so many are exhausted; we are always available. This availability has made our personal world a very loud and demanding place. What we need is a technological sabbatical, a time when we push away from all of our devices and just rest.

I made a reference earlier in this book to my time as a youth director hosting the annual summer hiking trip in the Porcupine Mountains Wilderness State Park in the Upper Peninsula of Michigan. For eight years we set aside a date in August and took a group of guys and girls for a period of four or five days. This trip consisted of hiking, sleeping in a tent, cooking over a fire, purifying our drinking water from the river, hanging our food up on bear poles and individual campsites several miles into the backcountry. For those of you who know me, you know that I am not a natural outdoors type. Interestingly enough, I fell in love with this activity for several reasons. The one-on-one time with the young men I camped with over the years proved to be a very special, influential, and spiritual time. This became a key time for building relationships and challenging teens spiritually during devotions. With the elimination of the many distractions of everyday life, focusing on the basics was easy.

When it was time to hike out of the woods after several days, it always amazed me how clear my thoughts and focus had become. There were times I entered this trip with heavy burdens, several important decisions I needed to make, and a mind filled with so many things I needed to begin working on. While driving back home, I always felt like everything had become calm and that the decisions I needed to make that had seemed so complicated before had become simple and obvious. I attributed this fresh feeling and clarity of thought to the fact that we had been in a place where we could be quiet for a while – a place where there was no cell service, no computers, no internet, and no way to communicate with the outside world. After a few days into this activity, we would often joke that the whole world could be falling apart and we wouldn't have a clue. The power and value of silence from technology quieted our minds and produced a peace that is highly needed by so many people today. Sadly, this peace is experienced far less than it could be.

Quiet time, a time away from technology, is time to remember what is really important. Family is important. Relationships and friendships are important. Reaching lost souls is important. Making time to serve God is important. All of these things take time, and that time is frequently stolen by the devices of our age.

See then that ye walk circumspectly, not as fools, but as wise, Redeeming the time, because the days are evil. – Ephesians 5:15, 16

So many things vie for our attention. Make sure that you do not allow electronics to rob you of spending time doing the things that are most important. Before you know it, your opportunity will be gone. Benjamin Franklin said, "Do not squander time for that is the stuff life is made of." You only have so much time. Use it well.

I must work the works of him that sent me, while it is day: the night cometh, when no man can work. – John 9:4

7 BIBLICAL PRINCIPLES OFTEN VIOLATED ON FACEBOOK

The principle of hearing both sides of a matter before you answer it

"He that answereth a matter before he heareth it, it is folly and shame unto him."
- Proverbs 18:13

Comments could be true or false, but opinions are formed without ever

getting both sides of a story.

The principle of concealment

"It is the glory of God to conceal a thing..." - Proverbs 25:2

"A talebearer revealeth secrets: but he that is of a faithful spirit concealeth the matter."
- Proverbs 11:13

Many times things that were personal matters get posted on a public

forum.

The principle of meddling

"He that passeth by, and meddleth with strife belonging not to him, is like one that taketh a dog by the ears."
- Proverbs 26:17

"But let none of you suffer as a murderer, or as a thief, or as an evildoer, or as a busybody in other men's matters."
- 1 Peter 4:15

There is the temptation for you to get involved in things that are none of

your business.

The principle of being wise concerning good and simple concerning evil

"...but yet I would have you wise unto that which is good, and simple concerning evil."
- Romans 16:19

Being exposed to –

Gossip

Slander

Profanity

Many times people have a boldness to say things online that they would

never say face to face.

"Be not deceived: evil communications corrupt good manners."
- 1 Corinthians 15:33

The principle of setting no wicked thing before your eyes

*"I will set no wicked thing before mine eyes: I hate the work of them
that turn aside; it shall not cleave to me."*
- Psalm 101:3

*"I made a covenant with mine eyes; why then should I think upon a
maid?"*
- Job 31:1

"Mine eye affecteth mine heart..."
- Lamentations 3:51

The Principle of redeeming the time

"Redeeming the time, because the days are evil."
- Ephesians 5:16

*"So teach us to number our days, that we may apply our hearts unto
wisdom."*
- Psalm 90:12

Secular psychologists have said that social networking is addictive.

Hours of time are wasted spent at a computer screen.

The principle of separating from backslidden or disgruntled brethren

> *"But now I have written unto you not to keep company, if any man that is called a brother be a fornicator, or covetous, or an idolater, or a railer, or a drunkard, or an extortioner; with such an one no not to eat."*
> - 1 Corinthians 5:11

> *"Now we command you, brethren, in the name of our Lord Jesus Christ, that ye withdraw yourselves from every brother that walketh disorderly, and not after the tradition which he received of us."*
> - 2 Thessalonians 3:6

People can leave a church and criticize, slander and give their opinion for all who want to listen at the click of a mouse! If we choose to stay in contact with them, we are violating Scripture and are at risk of our own heart being changed.

Pastor Pat Burke
Victory Baptist Church
Fredonia, NY

RUIN YOUR TESTIMONY
A HOW-TO GUIDE FOR USING SOCIAL MEDIA THE WRONG WAY

Be a liar. Are you saying or spreading things about people that aren't true?
"Thou shalt not bear false witness against thy neighbour."- Exodus 20:16

Be a fake. Are you a Christian at church, but not on Facebook?
"A double minded man is unstable in all his ways." – James 1:8

Be a jerk. Are you using social media to hurt others instead of help?
"But I say unto you, Love your enemies, bless them that curse you, do good to them that hate you, and pray for them which despitefully use you, and persecute you;" – Matthew 5:44

Be a loser. Are you spending more time on the internet than you are working?
"For even when we were with you, this we commanded you: that if any would not work, neither should he eat." – 2 Thessalonians 3:10

Be a thief. Are you illegally downloading music, movies or programs?

"Thou shalt not steal." – Exodus 20:15

Be a pervert. Are you looking at/for things you know you shouldn't?
"...That whosoever looketh on a woman to lust after her hath committed adultery with her already in his heart." – Matthew 5:28

Be a punk. Are you disrespectful to your parents or other authority?
"Honour thy father and they mother: that thy days may be long upon the land..." - Exodus 20:12

Be a reject. Are you disregarding what's happening in church to focus on your Facebook page?
"Remember the Sabbath day, to keep it holy." - Exodus 20:8

Bro. Tim Bish
Media & Communications Department
Grace Baptist College

10 TECHNOLOGY

Christian living and the use of technology is more than a list of do's and don'ts —it's a way of life. Everything a Christian does should bring glory to God (1 Corinthians 10:31). Using the digital world of technology is no different.

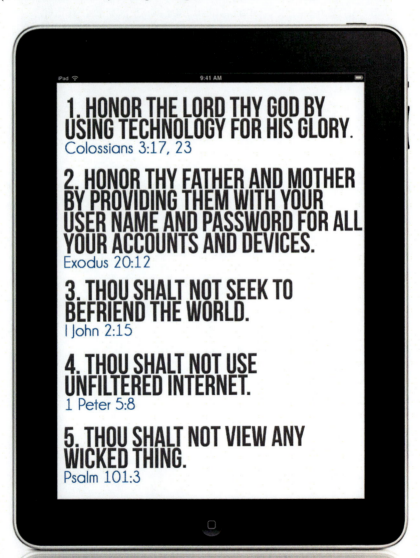

COMMANDMENTS

If a Christian is to avoid defiling himself with the use of technology, he must allow himself to be shielded from evil and guided by Biblical, principled living. These 10 Technology Commandments are helpful Biblical guidelines.

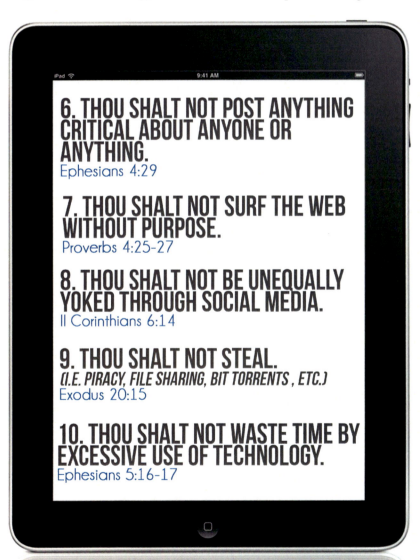

6. THOU SHALT NOT POST ANYTHING CRITICAL ABOUT ANYONE OR ANYTHING.
Ephesians 4:29

7. THOU SHALT NOT SURF THE WEB WITHOUT PURPOSE.
Proverbs 4:25-27

8. THOU SHALT NOT BE UNEQUALLY YOKED THROUGH SOCIAL MEDIA.
II Corinthians 6:14

9. THOU SHALT NOT STEAL.
(I.E. PIRACY, FILE SHARING, BIT TORRENTS , ETC.)
Exodus 20:15

10. THOU SHALT NOT WASTE TIME BY EXCESSIVE USE OF TECHNOLOGY.
Ephesians 5:16-17

4 QUESTIONS TO ASK
WHEN VIEWING WEBSITES, PARTICIPATING IN SOCIAL MEDIA, AND BROWSING VIDEO SITES

1. Would God be pleased with what I am viewing and browsing?

2. Will this activity hurt my personal Christian testimony?

3. Will this action hurt me spiritually?

4. Would I be ashamed or embarrassed if my parents, pastor, or youth director were standing over my shoulder watching what I am doing?

DID YOU KNOW?
INTERNET & TECHNOLOGY STATISTICS

- YouTube scans 100 years' worth of video every day.
- 1 in 6 marriages happen because the couple met online.
- The entire World Wide Web was controlled by 1 computer in 1989.
- 81% of all emails are spam.
- The internet is about 4 exabytes of information. 1 exabyte = 1,000 petabytes; 1 petabyte = 1,000 terabytes; 1 terabyte = 1,000 gigabytes! This number doubles every year.
- Yahoo! was originally named "Jerry and David's Guide to the World Wide Web."
- At least 2 people have been murdered for unfriending someone on Facebook.
- The first email was sent in 1971.
- Gmail.com was once owned by the Garfield cartoon.
- The very first website, created in 1993, is still up and running.
- There is an average of 2,600 Tweets per second.
- The first webcam was used to monitor a pot of coffee.
- The iPhone 4 is about 2,000 times faster than the Super Nintendo.
- The average 21 year old has played over 5,000 hours of video games, exchanged over 250,000 emails, text or instant messages, and has spent over 10,000 hours using a mobile phone.
- Asteroids and Lunar Lander were the first video games copyrighted in the US in 1980.
- Amazon's yearly revenue is greater than half of the world's Gross Domestic Products.
- More people have mobile cell service than have access to safe drinking water.
- 200 million+ tablets will be sold in 2013.
- 14.3 trillion web pages are live on the internet.
- Smartphone users have been found to be more productive than typical cell phone users.
- The average social network user spends 3.2 hours each day on networking sites.
- Facebook users combined spend 20,000 years' worth of time on the network each day.
- 8,500 likes are given on Instagram every second.
- 40 million new photos are taken with Instagram every day.

CYBERBULLYING

SOCIAL MEDIA BECOMES THE NEW SCHOOL YARD FOR BULLIES

Teens say cruel behavior takes place on...

TWITTER
23.8%

FACEBOOK
92.6%

MYSPACE
17.7%

INSTANT MESSENGER
15.2%

WHEN BEING BULLIED

65.8%
of teens responded to the bully
(35% responding in person)

15.4%
avoided school

4.5%
have been in a physical
fight with their bully

PARENTS REMAIN OBLIVIOUS

25%
of teens claimed
to be targets
of cyberbullying

2/3
of all teens have
witnessed cruel
behavior online

10%
of parents are aware
their teens are targets
of cyberbullying

Source: TRU Insights & McAfee, May 2012

SOCIAL MEDIA STATS & FACTS OF 2013

TWO FACTORS DRIVING SOCIAL MEDIA IN 2013

Mobile –
with the number of people accessing the internet via a mobile phone increasing by 60.3% to 818.4 million in the last 2 years.

60.3%

79%

Older Users Adoption –
On Twitter the ages 55-64 are the fastest growing, with 79% growth rate since 2012. On G+ and Facebook ages 45-54 are the fastest growing, with 46% and 56% growth rate.

f FACEBOOK

1 BILLION

Facebook **continues to grow** and make money via ads and mobile users. The latest facts and figures from its earnings call for the first quarter of 2013.

Daily active users — **665M**
Monthly mobile users — **751M**
Monthly active users — **1.1B**

% Mobile Ad Revenue

23% — Mobile ads (2012)
30% — Mobile ads (2013 Q1)

TWITTER

Twitter is the **fastest** growing social network in the world by active users. Twitter's fastest growing age demographic **is 55 to 64 year olds,** registering an increase in active users of 79%.

Monthly active users **288M**

44%
Growth from June 2012 - March 2013

21%
Twitter has 288 million monthly active users. That means that 21% of the world's internet population are using Twitter every month.

TWEET!

YOUTUBE

YouTube reaches more U.S. adults ages 18-34 **than any cable network.** Hours of video watched **DOUBLED** from last May when only 3B hrs were watched.

Unique monthly visitors **1B**
Hours of video watched a month **6B**

GOOGLE PLUS

Google+ is making an impact on the social media universe and is now the **second largest** social network.

Monthly active users **359M**

G+ active user base grew
33%
from June 2012 - March 2013

in LINKEDIN

The **largest professional** business network on the planet continues to grow but not at the pace of Twitter or Google+.

Users **200M+**

00:01:00
2 new users join LinkedIn every second

64% of the users are outside the USA

INFOGRAPHIC BROUGHT TO YOU BY:
growing social media   **surgogroup**

5/21/2013
SOURCES:
GlobalWebIndex Study • Visual.ly • Gigaom • www.jeffbullas.com

SPIRITUAL ADVISORY

WARNING:

This device has the ability to allow you to see and hear things that could potentially hurt you spiritually. Without proper caution and strict adherence to biblical principles, the user could be in extreme spiritual danger. DO NOT proceed with caution, but proceed only under biblical principle and through the proper spiritual filtering system found in Philippians 4:8.

A Publication Of Grace Baptist College

LIVING BY BIBLICAL PRINCIPLE
IN A DIGITAL WORLD

A Ministry Of Grace Baptist Church

ABOUT THE AUTHOR

Dr. Derek Hagland, Vice President of Grace Baptist College, Gaylord, Michigan, has worked as an assistant to Dr. Jon M. Jenkins since August of 1998. In his early years on staff at Grace Baptist Church, he served for over a decade as the youth pastor. During this time he led the church's annual Teen Spectacular youth conference, ran the church's teen church, and taught in the Christian school. In 2009, Dr. Hagland assumed the position of college Vice President and in 2011 was awarded an honorary doctorate  from Dr. Jenkins and the Grace Baptist College. In addition to his college responsibilities, Dr. Hagland currently hosts a daily radio broadcast for teens and travels to speak in camps, conferences, and revivals.